Book 1
Robotics
By Kenneth Fraser

&

Book 2
Human-Computer Interaction
By Solis Tech

&

Book 3
Raspberry Pi 2
By Solis Tech

Book 1
Robotics

By Kenneth Fraser

The Beginner's Guide to Robotic Building, Technology, Mechanics, and Processes!

Robotics: The Beginner's Guide to Robotic Building, Technology, Mechanics, and Processes!

Robotics: The Beginner's Guide to Robotic Building, Technology, Mechanics, and Processes!

Table of Contents

Introduction..5

Chapter 1 - What is a Robot?..6

Chapter 2 - Getting Started...10

Chapter 3 - Understanding Actuators...15

Chapter 4 - Microcontrollers and Motor Controllers............................19

Chapter 5 - Controlling Your Robot and Use of Sensors.......................22

Chapter 6 - Assembling and Programming a Robot 25

Conclusion...30

Introduction

I want to thank you and congratulate you for purchasing the book, *"Robotics: The Beginner's Guide to Robotic Building"*.

This book contains proven steps and strategies on how to build your own robot that will perform certain functions as you want it to do.

For most people, a robot is a machine that could mimic a human such as R2D2 and C3PO in Star Wars. But these types of robots are still in the figments of our imaginations. We are still far from giving robots high level of artificial intelligence to easily adapt and interact to its environment. There is however pioneering works on artificial intelligence that hopes to create humanoid robots.

The type of robots that exist and working today are robots that are programmed to do things that are too dangerous for humans, too repetitive, or just plain messy. These robots are often found in wide range of industries and places such as oil refineries, hospitals, laboratories, factories, and even in the Outer Space. There are about more than a million robots are working in different fields today.

There are types of robots that bring joy to kids such as the popular AIBO ERS-220 that is a bestseller toy during Christmas. While some robots perform great feats by discovering new places and gathering important data in the name of science, specialized robots such as the Mars Rover Sojourner and the underwater robot Caribou are sent to places that average humans cannot go.

Robots are exciting machines to play with, but they are more exciting to build. For hobbyists, building their own robots that capable of doing whatever they program these machines to do gives them pure delight.

This book introduces you to the science of robotics – its basic elements and fundamental concepts. And at the course of your reading, you will learn all the essential aspects you need to build your own robot.

Thanks again for purchasing this book, I hope you enjoy it!

Chapter 1 – What is a Robot?

It is interesting to know that even with all the hype about robots and with all the milestones in robotics, there is still not standard definition for a robot. There are, however, some basic characteristics that a robot should have and this could help you determine if a certain object is a robot or not. It will also guide you in deciding what features you need to build into a machine before you can say that it is a robot.

Four Basic Characteristics of a Robot

A robot has four basic characteristics: sensing, movement, intelligence, and energy.

Sensing

Basically, a robot must be capable of sensing its environment much similar to the way humans sense its surroundings. Robots could either sense through light sensors that mimic the functions of the eyes, or be equipped with chemical sensors that function like the nose, sonar sensors like the nose, touch sensors like the skin, and taste sensors like the tongue. These sensors will help the robot to become aware and understand its environment.

Movement

A robot should have the ability to move around through walking on legs, rolling on wheels, or through propellers. It's either the entire robot is able to move or just some parts of it such as head, arms, or just legs.

Intelligence

A robot should be equipped with artificial intelligence or AI. This is usually done through computer programing. Hence, you need a background in programming to provide your robot with the needed intelligence. You need to program the robot's intelligence so that it will know what to sense and how to move.

Energy

A robot should have a way to power itself. The energy source could be electrical, chemical (battery), or solar. The method by which your robot energizes itself depends on what your robot is required to do.

Robotics: The Beginner's Guide to Robotic Building, Technology, Mechanics, and Processes!

Working Definition of a Robot

For the purpose of discussion and for reference, we define robot as a machine that contains control systems, sensors, manipulators, software and power supplies that works together to do certain tasks.

Building a robot requires understanding of the fundamental principles of mechanical engineering, mathematics, physics, and computer programming. In special cases, it also requires specific knowledge on chemistry, biology, and medicine. In studying robotics, you need to be actively engaged with wide range of disciplines to build robots that could solve certain problems.

A Brief History of Robotics

The word "robot" was first used in a play entitled R.U.R (Rossum's Universal Robots) written in 1921 by Czech writer Karl Capek. This play is about machines that are built to work on a factory and eventually revolted against their human masters. Robots are the Czech word for slave.

Meanwhile, the word robotics also first appeared in a work of fiction. Russian-born American fictionist Isaac Asimov used it in his short story "Runabout" (1942). Compared to Capek, Asimov had a more positive opinion of the role of robots in the society. In general, he described robots as useful machines that serve humans and perceived them as a "better, cleaner race. He also proposed the three Laws of Robotics:

First Law of Robotics

A robot may not injure a human or, through inaction, allow a human to come to harm.

Second Law of Robotics

A robot must obey the orders given by humans except if such orders would violate the First Law.

Third Law of Robotics

A robot may protect itself as long as such protection do not violate with the First Law or Second Law of Robotics.

Early Models of Robots

Among the earliest cases of a mechanical system designed to perform a regular task was recorded around 3000 BCE. Egyptian water clocks are added with human statuettes to hit the hour bells and signal the passing of time. In 400 BCE,

Archytus of Taremtum, who was known as the inventor of pulley and screw, created a pigeon made of wood that is capable of flying. Meanwhile, hydraulically-powered figurines that could speak prophecies were common during the Greek domination of Egypt during the second century BCE.

In the first century C.E., Petronius Arbiter built a doll that is capable of moving like a human being. In 1557, Giovanni Torriani built a wooden robot, which could fetch the Emperor's bread every day from the store. By 1700s, robotic inventions became common with numerous impractical yet ingenious machines such as steam-powered automata crafted in Canada as well as the popular talking doll by Thomas Edison. Even though these creations may have inspired the design and functions for the modern robot, the progress during the 20th century in the field of robotics exceeded previous advancements many times over.

The First Modern Robots

The robots that we are familiar with were built by George C. Devol in the 1950s. The inventor from Kentucky designed and patented a reprogrammable manipulator that he dubbed as "Unimate" derived from "Universal Automation." For years, he tried commercializing his product, but failed. But in 1960s, the entrepreneur-engineer Joseph Englberger bought the patent from Devol and modified it into an industrial robot. He established his company, Unimation, for production and marketing of these products. He was successful in this venture, and in fact, Englberger is regarded today as the Father of Robotics.

Robotics also progressed within the academic institutions. Alan Turing, pioneering computer scientist, mathematician, logician, and cryptologist, published his book "Computing Machiner and Intelligence" where he proposed a test to determine if a machine has the capacity to think for itself. This test is known as the Turing Test.

In 1958, Charles Rosen of the Stanford Research Institute created a research team to work in the development or a robot known as "Shakey" that was more advanced compared to Devol's Unimate. Shakey can move around through the room, sense light through his "eyes" move around strange environment, and to a particular degree, and react to what is happening to his surroundings. He was called Shakey because of his clattering and rickety motions.

In 1966 at Massachusetts Institute of Technology (MIT), Joseph Weizenbaum created an artificial program named ELIZA, which functions as a computer psychologist that manipulates its user's statements to formulate questions.

In 1967, Richard Greenblatt developed MacHack, a program that is capable of playing chess, as a response to a critical article written by Hubert Dreyfuss where he boasted that no computer program can beat him in chess. When the program is finished, Dreyfuss was invited to play and was defeated. This program was the foundation of future chess programs that eventually developed into Big Blue, the program that defeated Grand Master Gary Kasparov in 1997.

The interest in robotics is one of the major catalysts in the development of computers. In 1964, the IBM 360 becomes the first computer to be produced massively.

Robots are also crucial in pioneering space explorations. In 1969, the United States successfully used the latest technology in robotics and computing for Neil Armstrong's landing on the moon. Robots also helped in the expansion of scientific knowledge. In 1994, Carnegie Universities crafted Dante II, an eight-legged walking robot that successfully descends into the crater of Mt Spur to gather samples of volcanic gas.

Commercial companies also leveraged on the mass appeal of robots. In 1999, Sony released its original version of AIBO, a robotic dog that can entertain, learn, and communicate with its owner. Advanced versions have followed in the succeeding years, with the final model, the ERS-7M3, released in 2005.

Honda also released its ASIMO robot, an advanced humanoid robot in 2000. In 2004, Epsom was hailed as the world's smallest robot (7 cm high and weighs only 10 g.) The robotic helicopter is designed to fly and capture videos during natural disasters.

After being released in 2002, a robotic vacuum cleaner known as the Roomba became a huge hit. It sold more than 2.5 million units, which shows that there's really a huge demand for domestic robot technology.

Hundreds of films feature robots such as The Day the Earth Stood Still (1951), Arthur C. Clark's 2001: A Space Odyssey (1968), Star Wars (1977), Blade Runner (1982), Terminator (1984), Nemesis (1992), I, Robot (2004), Transformers (2007), and many more. The popularity of robotic films shows that people are inspired and delighted by the idea of machines that can independently move and think for itself.

If you are ready to build your own robot, continue to the next Chapter to help you get started.

Chapter 2 – Get Started

The first step in building your own robot is to determine what it should do, that is, your purpose of why you are building the robot. Robots can be used in different situations and are mainly designed to assist humans. It will help you a lot to learn first the different purposes and uses of robots.

Basically, robots are divided into two main groups: industrial and domestic robots.

Industrial Robots

Industrial robots are used in factories to manufacture products with precisions such as computers, cars, cellphones, medicine, and even food. Robots increased the productivity in different workplaces, which resulted to booming industries. Each type of industrial robot has its specific form that corresponds to its function. For instance, robotic arms are often used in car assembly lines to spray paint or weld frames. Robotic arms are among the most common robots today. Recently, agricultural robots have been introduced mainly to perform farm tasks such as cutting weeds and harvesting crops.

Domestic Robots

Domestic robots are mainly used in the home to perform household chores. They usually perform repetitive tasks every day such as vacuuming floors, mowing the lawn, vacuuming floors, and other chores that people usually don't have time to do. For example, there are vacuum robots that can clean the floors. They are equipped with motion sensors so they will not run into any object. You just need to push the switch on and it will do its job. It could pick up dust and pet hairs and could be used for hours.

There are also mower robots that could mow lawns. They are equipped with sensors to detect grass edges. Domestic robots are also used for entertainment such as Robosapien, AIBO, and iDog.

Choosing a Robotic Platform

The next step in building your own robot is to decide on the type of robot you want to build. A usual robot design usually begins with "inspiration" of what the robot will do and what it will look like.

The types of robots that you can build are endless. As long as you can envision something that a robot can do, you can work your way to achieve it. But for beginners, you can start with the following types: land robots, aerial robots, aquatic robots, stationary robots, and hybrid robots.

Robotics: The Beginner's Guide to Robotic Building, Technology, Mechanics, and Processes!

Land Robots

Land-based robots, particularly those added with wheels are among the most common mobile robots built by beginners, because they often require minimal investment while providing the opportunity to learn more about robotics. Meanwhile, the most advanced type of robot is the humanoid robot, which is akin to humans. Humanoids require several degrees of freedom and synchronization of different motors and use several sensors.

Wheeled Robots

Wheels are among the most common method of adding mobility to a robot and are used to mobilize many different sizes of robots and robotic platforms. Wheels could be about any size, and there's no limit in the number of wheels that you can add. More often than not, robots that are equipped with three wheels are using two wheels and a caster at one end. More advanced robots with two wheels are using gyroscopic stabilizing technology.

Meanwhile, robots that are added with four to six wheels usually use several drive motors that decreases the risk of slippage. Also, mecanum wheels or omni-directional wheels can provide the robot considerable benefits in mobility. Most beginners in robotic building are mistaken in thinking that inexpensive DC motors can mobilize robots that are medium in size. As you will learn later, there are more factors that you need to consider before you can add mobility to your robot.

Advantages

Wheeled robots are ideal for beginners as they are often more affordable to build. They have simple design and construction, and there are unlimited options. In addition, robots with six wheels or more could rival the mobility of a track system.

Disadvantages

Wheeled robots usually have small contact area, because only a small portion of the wheel is touching the ground. This results to lower traction that may cause slippage.

Tracked Robots

Tracks are used in tanks for mobility. Even though tracks, also known as treads, don't provide the added torque, they can decrease slippage and can equally distribute the robot's weight. This makes the robot easier to mobilize in loose ground such as gravel and sand. In addition, flexible track systems could easily navigate through a bumpy surface. Most hobbyists also believe that tank tracks are quite cool compared to wheels.

Advantages

Steady contact with the ground avoids slippage, which is prevalent with wheels. The track system also distributes weight evenly, which helps the robot in navigating different surfaces. Tracks can also be used to extensively enhance the ground clearance of the robot without adding a bigger drive wheel.

Disadvantages

The main disadvantage of using a track system for robots is that in turning, there's the tendency to cause damage to the surface that also causes damage to the tracks. In addition, robots are often built around the tracks, and there's a limit in the availability of the tracks. Drive sprocket can also considerably restrict the number of motors that you can use.

Legged Robots

More and more robots are using legs for movement. Legs are usually ideal to use for robots that should navigate on uneven ground. Many prototype robots are built with six legs that allow the robot for static balance. Robots with fewer legs are more difficult to balance as it requires dynamic stability. Once the robot ceases moving in the middle of the stride, it could fall over. Even though there were robots with one leg moving by hopping, bipeds, quadrupeds, and hexapods are the most common forms.

Advantages

The leg motion is the most natural among the platforms, and it can easily overcome big obstacles and move through rough surface.

Disadvantages

Most beginners are discouraged in building their first robot that moves using legs, as it requires high level of electronic, mechanical and coding skills. You also need to find a small battery that can provide the required power, so legged robots are usually expensive to build.

Aerial Robots

Humans have long been inspired by the idea of flight, and this transcends into the field of robotics. The idea of Autonomous Unmanned Aerial Vehicle (AUAV) has gained popularity over the years, and many enthusiasts have developed numerous prototypes. However, the benefits of crafting aerial robots have yet to prevail over the disadvantages. In building aerial robots, many hobbyists are still using commercial remote controllers. Professional aircrafts such as the Predator commissioned by the US military were partially autonomous though recently, updated versions of the Predator have completed aerial missions with only minimum human intervention.

Advantages

Aerial robots are great for surveillance, and remote controlled aircraft has been developed through the years, so there is a diverse community for mechanics where you can find support and know-how in building your own aerial robots.

Disadvantages

There is still limited community when it comes to autonomous control, as most of the knowledge on this field is protected by the US military. Meanwhile, this robot type is expensive as the whole robot could be broken if you miscalculate the steps and lead the robot into a crash.

Aquatic Robots

Recently, more and more hobbyists, communities, and companies are building unnamed aquatic vehicles. There are still many hindrances to overcome in order to make aquatic robots more enticing for the wider communities in robotics. But it is interesting to take note that there are companies today who are manufacturing robots that can clean pools. Aquatic robots can use thrusters, ballast, wings, tails, and fins to move under water.

Advantages

A massive part of the ocean is still unexplored so there's a lot to discover if you choose to build aquatic robots that could help in discovering the underwater world. The robot design is also guaranteed to be unique, and it could be tested in a pool.

Disadvantages

Aquatic robots are often very expensive to build, and there is the risk that the robot could be lost while deep in the ocean. You should also take note that most electronic parts don't pair well with water, especially salty water. You also need to consider the water pressure as going beyond deep sea needs considerable investment and research. There is also very limited robotic community that can provide support, and also limited wireless communication options.

Hybrid Robots

Your concept for the robot may not easily fall into any of the categories mentioned above or could be composed of various functional components. Take note that this book is written to guide you in building mobile robots and not those with fixed designs. In building a hybrid design, it is best to use a modular design where each functional component could be taken off and tested as a separate part.

Advantages

Hybrid robots are designed and built according to your preferences and needs. These robots could be used for various tasks and can be composed of modules. Hybrid robots could lead to versatility and increased functionality.

Disadvantages

Hybrid robots are often complicated to build and expensive. Parts need to be customized to fit the design.

Grippers and Arms

Even though grippers and arms don't fall under the category of mobile robots, robotics basically began with end-effectors and arms. Grippers and arms are the most ideal way for a robot to interact with the environment it is dealing with. Basic robotic arms could have just two to three motions; while more advanced arms could have more than a dozen movements.

Advantages

Most robotic arms and grippers have simple designs, and it is easy to make a three to four degree of freedom robotic arm with a turning base and two joints.

Disadvantages

Robotic arms are stationary unless you fix them on a mobile platform. The cost of building arms or grippers depends on the lifting capacity you need.

In the next chapter, you will learn how to choose the right actuators or motors for your robot.

Chapter 3 – Understanding Actuators

After learning general information about robots and robotics in the first two chapters, it is now time to choose the right actuators to mobilize your robot.

What Are Actuators?

Actuators are devices that transform energy into physical motion. In robotics, this energy is usually electrical energy. Most actuators today produce either linear or rotational motion. For example, a DC motor is a type of actuator.

Selecting the right actuator for your robot requires learning the available actuators, and some fundamental knowledge of physics and mathematics.

Rotational Actuators

Rotational actuators convert electrical energy into rotating motion. There are two primary mechanical parameters that distinguish each actuator: (a) the rotational speed that is often measured in revolutions per minute or rpm and (b) torque or the force that the devices can produce at a given distance often expressed in Oz•in or N•m.

AC Motor

Alternating Current (AC) is rarely used in robots because most of them are powered through Direct Current (DC) in form of cells or batteries. In addition, electronic parts use DC, so it is easier to use the same type of power supply for the actuators. AC motors are primary used in industrial settings where high torque is necessary or where the motors are connected to a wall outlet.

DC Motor

DC motors are often cylindrical in shape but they also come in different shapes and sizes. They also have output shafts that rotate at high speed often between 5000 and 10000 rpm. Even though DC motors rotate very fast, most have low torque. To decrease the speed and add torque, a gear could be added. To install a motor into a robot, you must fix the body of the motor to the robot's frame. Hence, motors usually have mounting holes that are basically located on the motor's face so that they can be easily installed. DC motors could either rotate in counter clockwise or clockwise. The angular movement of the turning shaft could be measured using potentiometers and encoders.

Geared DC Motor

A DC Motor could be added with a gearbox to reduce the motor's speed and enhance its torque. For instance, if a DC motor rotates at 5000 rpm and produces a 0.0005 N•m of torque, adding a 123:1 ("one hundred and twenty three to one") gear would reduce the speed by a factor of 123 (resulting to 5000 rpm / 123 = 40 rpm) and increase the torque by a factor of 123 (0.0005 x 123 = 0.0615 N•m). The most common types of gears are planetary, spur, and worm. Similar to a DC motor, a geared DC motor can also rotate in either clockwise or counter clockwise. You can add an encoder to the shaft if you want to know the number of rotations of the motor.

Hobby Servo Motors

Hobby Servo Motors, also known as R/C Servo Motors are actuators that rotate to a certain angular position, and were traditionally used in more expensive remote controlled machines for controlling or steering flight surfaces. Today, they are used in different applications so their prices have been reduced considerably, and the variety has also increased. Most servo motors can only rotate about 180 degrees. A hobby servo motor is composed of a DC motor, electronics, gears, and a potentiometer that measures the angle. The latter works with the electronics to mobilize the motor and stop the output shaft at a certain angle. In general, these servos have three wires, voltage in, control pulse, and ground. A robot servo is a recently developed servo that provides both position feedback and continuous rotation. Servos could rotate clockwise or counterclockwise.

Stepper Motors

As the name implies, stepper motors rotates following certain steps or degrees. The number of degrees the shaft rotates with every step could vary depending on various factors. Majority of stepper motors don't include gears, so similar to a DC motor, the torque is quote low. Fixing gears to a stepper motor has similar effect as installing gears to a DC motor.

Linear Actuators

Linear actuators produce linear movements. They have three primary distinctive mechanical properties: (a) the force measured in kg or lbs (b) speed measured in m/s or inch/s and (c) the maximum and minimum distance that the rod could move also known as the stroke measured in inches or mm.

Linear DC Actuator

A linear DC actuator is usually composed of a DC motor attached to a lead screw, which also turns as the motor moves. The lead screw has a traveler that is forced either away or towards the motor, basically transforming the rotating motion to a linear movement. Some DC linear actuators integrate a linear potentiometer that adds a linear position feedback.

Solenoids

Solenoids are comprised of a coil wound surrounding the mobile core. Once the coil is energized, the core is forced away from the magnetic field and creates a motion in one direction. Several coils or some mechanical arrangements will be needed to provide movements in different directions. A solenoid stroke is often very small but they are often very fast. The strength primarily depends on the size of the coil and the electrical power passing through it.

Hydraulic and Pneumatic Actuators

Hydraulic and pneumatic actuators use liquid or air respectively to create a linear movement. These actuators could have lengthy strokes, high speed and high force. To use these actuators, you need to use a fluid or air compressor that makes them harder to use compared to basic electrical actuators. These are often used in industrial applications because of their large size and high force speed.

Muscle Wire

Muscle wire is a specialized wire, which contracts when electricity passes through it. When electricity is gone and once the wire cools down, it will go back to its original length. This type of actuator is not fast, strong, or creates a long stroke. Nonetheless, it is one of the most convenient actuators to use if you need to work with smaller parts.

How to Choose the Proper Actuator for Your Robot

To guide you in choosing the actuator for certain tasks, consider answering the following questions to help you.

Take note that new innovations and technologies are always being released regularly, so nothing is permanent. Also remember that one actuator could perform various tasks in various contexts.

1. Do you need to mobilize a wheeled robot?

Drive motors should carry the weight of the whole robot and will most likely need a gear down. Majority of the robots utilize "skid steering" while trucks or cars utilize rack and pinion steering. If you prefer the skid steering, geared DC motors are recommended to use for robots with tracks or wheels. Geared motors provide constant rotation, and could have discretionary position feedback through optical encoders. Because the rotation needed is limited to a certain angle, you can choose a hobby servo motor for stirring.

2. Is there a limit on the range of motion?

If the range is restricted to 180 degrees and the needed torque is not a critical factor, a hobby servo motor is recommended. Servo motors are available in various torques and sizes and comes with angular position feedback. Majority of these motors use potentiometer, while some specialized ones use optical encoders. R/C servos are now popularly used to build small walking robots.

3. Do you need a motor to lift or turn heavy loads?

Raising a weight needs considerably more power compare to moving a weight on a flat surface. Torque should be prioritized than the speed, and it is ideal to use a gearbox with a powerful DC motor or a linear DC actuator with a high gear ratio. You can use an actuator system that could prevent the mass from falling if there is a disruption in the power source. This includes clamps or worm gears.

4. Do you need the angle to be precise?

Stepper motors that are paired with a motor controller cold provide a very precise angular motion. They are more ideal to use compared to servo motors because they provide constant rotation. But there are also high-end digital servo motors that use optical encoders and can provide high precision.

5. Do you need to achieve movements in a straight line?

Linear actuators are ideal for moving parts and placing them in a straight line. They are available in different configurations and sizes. For fast movements, you must consider solenoids or pneumatics, for high torques, you can use linear DC actuators or hydraulics, and if the movement requires minimum torque, you can use muscle wire.

Chapter 4 - Microcontrollers and Motor Controllers

Microcontrollers are considered as the "brain" of the robot because it is responsible for all decision making, computations, and communications. These are devices with the capacity to execute a program (a series of instructions).

To interact with the external world, a microcontroller has a sequence of electrical signal connections (known as pins), which could be switched on or off using programming functions. These pins are also used in reading electronic signals that are released by sensors or other devices and determine if they are low or high.

Majority of microcontrollers today could measure analogue voltage signals, or signals that could have a full range of values rather than just two specified states by using analog to digital converter or ADC. Through the use of ADC, a microcontroller could assign a numerical value to the analog voltage that is neither low nor high.

What Could Microcontrollers Do?

Numerous complicated actions could be achieved by setting the pins low and high creatively. Nonetheless, building complicated algorithms such as smart movements and data processing or complicated programs are not yet on the range of microcontrollers because of its natural speed and resource limitations.

For example, to light a blinker, you can program a repeating sequence in which the microcontrollers could turn a pin high, wait for several seconds, turn it low, wait for several seconds and goes back to the first sequence. A light that is connected to the pin will then blink open-endedly.

Similarly, microcontrollers could be used to take control of other electronic devices including actuators when they are installed to motor controllers, Bluetooth or WiFi interfaces, storage devices, and many more. Because of its versatility, microcontrollers could be found in common everyday products. Basically, every home electrical device or home appliance utilizes at least one microcontroller.

Not similar to microprocessors found in Central Processing Units in personal computers, microcontrollers don't need peripherals such as external storage devices or external RAM to operate. Hence, even if the microcontrollers are less powerful compared to microprocessors, building circuits and products based on microcontrollers is an easier task and a lot more affordable, because minimal hardware parts are needed. Remember, microcontrollers can output minimum amount of electrical power through pins. Hence, a generic microcontroller cannot power solenoids, power electrical motors, large lights, or other direct loads. Doing this could cause physical damage to the controller.

Programming Microcontrollers

There's no need to shy away from programming microcontrollers. Not similar in the past where making a blinker took comprehensive knowledge of microcontroller and at least a dozen line of code, programming microcontrollers is fairly easy today. You can use the simplified Integrated Development Environments (IDE), which uses modern languages, full line archives that could cover all of the most common actions, and several handy samples to help you get started. You can learn more about programming your robot in Chapter 6.

How to Choose the Proper Microcontroller for Your Robot

You will need a microcontroller for any robotic building project unless you're into BEAM robotics or you want to control your robot through an R/C system or a tether. For starters, selecting the right microcontroller could seem like a difficult job, particularly considering the product range, specifications, and applications. There are various microcontrollers available today such as BasicATOM, POB Technology, Pololu, Arduino, BasicX, and Parallax.

The following questions could guide you in choosing the right microcontroller:

1. Which microcontroller is widely used for your type of robotic project?

Building robots is not a popularity contest, but the fact that a microcontroller has a large supporting community or has been used in the same project can make the design phase easily. With this, you can benefit from other experience and hobbyists. It is common for hobbyists to share codes, pictures, instructions, and videos even lessons learned.

2. Do you need specific accessories for a certain microcontroller?

If your robot has special needs or there is a certain accessory or component that is important for your design, selecting a compatible microcontroller is clearly essential. Even though most accessories and sensors could be directly interfaced with most microcontrollers, some accessories are designed to interface with a particular microcontroller.

3. Do you need special features for your robot?

A microcontroller should be able to perform all the special actions needed for your robot to function well. Some features are common to all microcontrollers such as being able to execute basic mathematical operations, having digital inputs and outputs, and making decisions. Others may need certain hardware such as PWM, ADC, and communication protocol support. You must also consider pin counts, memory and speed requirements

Motor Controllers

Motor controllers are electronic devices that serve an intermediary device between a microcontroller, the motors, and the power supply.

Even though the microcontroller decides the direction and the speed of the motors, it doesn't have enough power to directly drive them. Meanwhile, the motor controller can supply the current at the needed voltage but doesn't have the capacity to decide how fast the motor must turn.

Hence, the microcontroller and the motor controller must work together to make the motors move accordingly. The microcontroller can provide instructions to the motor controller on how to power up the motors through a standard and basic communication method such as PWM and UART. In addition, some motor controllers could be manually regulated using an analog voltage often created through a potentiometer.

The size and weight of a motor controller may greatly vary from a device that is smaller than the tip of a pencil to a huge controller that could weight several kilos. The size and weight often has a minimum effect on the robot, unless you want to build unnamed aerial or aquatic robots.

Types of Motor Controllers

Because there are several types of actuators (as we have discussed in Chapter 3), there are also several types of motor controllers: brushed DC motor controllers, brushless DC motor controllers, servo motor controllers, and stepper motor controllers.

How to Choose a Motor Controller

You can only choose a motor controller after you have decided on what type of actuator you want to use. In addition, the current that a motor draws depends on the torque it could provide. A small DC motor will not use much current, but cannot also release much torque, while a bigger motor could release higher torque but will need increased current.

Chapter 5 - Controlling Your Robot and Use of Sensors

Based on our definition of a robot, it should gather data about its surroundings, make smart decisions and then execute actions based on calculations. This also includes the option for the robot to become semi-autonomous (with aspect that are controlled by humans and other aspects that it can do on its own).

One good example of this is a complex aquatic robot. A human controls the basic motions of the robot while an installed processor measures and reacts to the underwater currents to keep the robot in one position while still preventing a drift. A camera installed in the robot would send videos back to the human while the sensors could track the water pressure, temperature, and more. Once the communication line falters between the robot and the human, an autonomous program could take over to instruct the robot to reach for the surface.

In controlling your robot, you need to figure out the level of autonomy. You need to choose if you want the robot to be tethered, wireless, or autonomous.

Tethered

Direct Wired Control

The simplest way to control a vehicle is by using a handheld controller that is physically connected to a vehicle using a tether or a cable. Knobs, switches, joysticks, buttons, and levers on the controller will allow you to control the robot without the need to add sophisticated electronics. In this setting, the power source and motors could be directly connected with a switch to control the rotation. These machines often have no artificial intelligence and are regarded as remote controlled devices than robots.

Wired Computer Control

Another method is to integrate a microcontroller into the machine but still using a tether. Attaching the microcontroller to your computer's ports will allow you to control the actions using the keyboard, a joystick, a keypad, or other device. Adding a microcontroller to your robot project may also require programming how the robot will respond to the input.

Ethernet

Another way to use computer control is to use an Ethernet interface. A robot that is directly connected to a router can also be used for mobile robots. Building a robot, which can communicate through the internet could be sophisticated, and usually a wireless internet connection is more recommended.

Wireless

Infrared

You can ditch away cables and wires if you use infrared transmitters and receivers. This is often a great achievement for beginners. Infrared control needs "line of sight" to function. The receiver should have the ability to see the transmitter to receive the data. Infrared remote controls can be used to send commands to infrared receivers that are paired with microcontrollers that interpret these signals and control the actions of the robot.

Radio Frequency

Remote control units often use microcontrollers in the receiver and transmitter for data transmission through radio frequency. The receiver box usually has a printed circuit board (PCB) that includes a small servo motor controller and a receiving unit. RF communication needs a transmitter matched with a receiver or a transceiver. RF doesn't need clear line of sight and could also provide considerable distance. Basic RF devices could allow for data transfer between devices between long distances, and there's no limit to the range for more RF devices.

Bluetooth

Bluetooth is a type of Radio Frequency and follows certain protocols for sending and receiving data. Standard Bluetooth range is usually restricted to about 10 meters although it has the advantage of controlling the robot though Bluetooth-enabled devices including laptops, smartphones, and PDAs. Similar to RF, Bluetooth provides two-way communication.

WiFi

Recent development in wireless technology enables you to control a robot through the Internet. To build a WiFi robot, you must have a wireless router that is connected to the internet and a WiFi device on the robot. You can also use a device, which is enabled with TCP/IP with a wireless router.

Autonomous

High-level robots are autonomous. With recent developments, you can now use the microcontroller in its full potential and program it to respond to input from the sensors. Autonomous control may come in different types: restricted sensor feedback, pre-programmed with no feedback from the environment, and complex sensor feedback. Genuine "autonomous" control includes different sensors and code to allow the robot to figure out by itself the smartest action to be taken in any situation.

The most sophisticated methods of control presently used on autonomous robots are auditory and visual commands. For auditory control, a robot will react to the sound of the human's voice for instructions such as "get the ball" or "turn left." For visual command, a robot may look to an object to decide on what to do. Instructing a robot to turn to the right by showing a drawing of an arrow that is pointing to the right requires complicated programming. Even though these things are no longer impossible, they need a sophisticated level programming and usually hundreds of hours.

Not similar to humans, robots are not restricted to just sound, sight, smell, touch, and taste. Robots use different electromechanical sensors to understand and discover their surroundings. Mimicking a natural organism's senses is presently a great challenge, so developers and robotic builders are using alternatives to these natural senses.

Chapter 6 - Assembling and Programming a Robot

After learning all about the fundamental blocks in building a robot, the next stage is the designing and building of the frame that will keep all the components together and will provide your robot a definite look and shape.

Constructing the Frame

There's no fix method in creating a frame, because there is often a trade-off to be constructed. You may prefer to use a lightweight frame but you may need to use costly materials. You may like a strong or big chassis but you may realize it is expensive, hard, or heavy to produce. The frame could be complicated and may take some time to design and build.

Materials

There are different materials that you can use in creating a frame for your robot. As you try different materials to construct not only robots but other types of machines, you will also understand the advantages and the disadvantages as to which material is the most suitable for a specific project. The roster of suggested building materials below comprises only the more common one, and when you have tried several of these materials, you can start experimenting or blending some together.

Basic Construction Materials

Some of the most basic construction materials could be used to build good-quality frames. The cheapest materials is the cardboard that you can usually find for free and could easily be bent, cut, layered, or bent. You can construct a reinforced cardboard box that looks a lot nicer and is more proportional when it comes to the size of your robot. You can then paint it with glue or epoxy to make it stronger then add extra layer of paint.

Structural Flat Materials

For a more durable frame, you can use a standard structural material such as a sheet of plastic, metal, or wood. You just need to puncture some holes to connect the electronic components. A stronger piece of wood has a tendency to be heavy and thick, while a thin sheet of metal could be too flexible. You can attach components to both sides and the wood will still remain solid and intact.

If you're at the stage where you are ready to have an outsourced frame, the best option is to acquire the part precision cut through a water or laser jet. Hiring a third-party to produce a custom part is recommended only if you are completely sure of the dimensions, because the mistakes could be expensive. Companies that offer computer controlled cutting services may also provide different other services such as painting and bending.

3D Printing

Building a frame constructed from 3D printed panels is not always the most structurally sound option, primarily because it is built up in several layers. However, this process could produce complex and detailed shapes that could be impossible to build using other methods. One 3D printed component may contain all the important mounting points for all mechanical or electrical parts without compromising the robot's weight. In the past decade, the cost of 3D printing is quite expensive, but as it becomes popular, the price of producing the components is also expected to go down.

Assembling the Parts Together

With the available options for materials and methods, you can now start assembling the parts together. You can follow the steps below to build a simple, aesthetic, and structurally reliable robot frame.

1. Decide on the material you want to use.

2. Gather all the parts that your robot will need, both mechanical and electrical and measure them. In case you don't have all the components ready, you can refer to the dimensions that are often supplied by the manufacturers.

3. Think of and draw various designs for your frame. It's fine not to provide details.

4. Once you find the suitable design, be certain that the structure is reliable and all the parts would be supported in the frame.

5. Sketch every component of the robot on cardboard or paper at true scale. You can also draw the parts in the CAD software and print them.

6. Test the design in CAD and in actual setting using your paper prototype by test fitting every component and connection.

7. Measure the dimensions again and when you are completely certain that your design is right, begin cutting the frame into the material. Take note to measure two times and only cut once.

8. Test fit every part before assembling the frame if in case you need some changes.

9. Construct the frame using appropriate assembling materials such as glue, nails, screw, duct tape, or any appropriate binding tools that you prefer.

10. Fit all the parts into the frame and there you go, you have just built your robot!

Constructing the Robot Parts

The last step discussed above should be described further. In the past chapters, you have already chosen the electrical parts including the actuators, microcontroller, and motor controller. The next step is to construct them so they will work together.

In the following section, we'll use standard cable colors and terminal names that encompass common parts. You must rely on manuals and datasheets when you are working on your specific parts.

Attaching Motor Controllers to Motors

A geared DC motor or a linear DC actuator usually has two wires: black and red. Attach the black wire to the M- terminal on the DC motor controller and the red wire to the M+ terminal. Connecting the wires the other way around will only cause the motor to rotate in the opposite direction. Meanwhile, servo motors have three wires: red, black, and yellow. A servo motor controller comes with pins that are matching these wires so you can just plug it directly.

Attaching Microcontroller to Motor Controllers

Microcontrollers can communicate with motor controller in different ways: 12C, R/C, Serial, or PWM. Be sure to refer to the manual for each microcontroller for specific instructions on proper connection. Regardless of the method you choose, the microcontroller and the logic of the motor controller should share matching ground reference. This can be achieved by attaching the GND pins together. Meanwhile, a logic level shifter is needed if the devices don't share the same logic levels.

Attaching Batteries to a Microcontroller or a Motor Controller

Majority of the motor controllers available today have two screw terminals for the battery labels marked with B- and B+. If the batteries you got are provided with a connector and the controller comes with screw terminals, you could still search for a pairing connector with wires that you can attach to the screw terminal. If this is possible, you need to find another way to link the battery to the motor controller while you can still unplug the battery and link it to a charger. It's possible that not all the electrical and mechanical components you have selected for your robot could operate on a single voltage, and so may need several voltage regulation circuits or batteries.

If you are building a robot with a microcontroller, DC gear motors, and maybe servo motors, it's easy to see how a battery may not be able to power every component directly. Nevertheless, it is best to choose a battery that can directly power as many devices that you need. The battery with the largest capacity must be connected with the drive motors. For instance, if the motors you select are rated a nominal 12 volts, the primary battery must also be 12 volts. So you can use a regulator to energize a 5 volts microcontroller. LiPo and NiMH batteries are the top choices for small to medium robots. Select NiMH if you need cheaper batteries and LiPo if you need light weight batteries. Always take note that batteries are powerful devices that could easily burn your circuits if they are not properly connected. Always make sure that the polarity is correct and that your device could handle all the energy supplied by the battery. If you are not certain, never make assumptions.

Adding Electrical Parts to Frame

You can attach electrical components to your frame through different methods. Make certain that whatever method you use, don't conduct electricity. Usual methods include screws, hex spacers, Velcro, double-sided tape, cable ties, glue, and many more.

Programming Your Robot

Programming is often the last step in building your robot. If you have followed the steps described in the previous chapters, by now you have selected the electrical components such as actuators, microcontrollers, motor controllers, sensors, and more. At this point, you might have already constructed your robots and hopefully it looks something like you want it to be. But without the proper program, your robot is just a cool paperweight.

It requires another book to teach you robotic programming. Instead, this section will guide you on how to get started and what you should learn.

There are several programming languages that you can use to program the microcontroller that will serve as the brain of your robot. The following are the most common programming languages you can choose:

Assembly

This programming language is just a shy away from programming a full-pledged computer, and so it could be difficult to use. This language is ideal to use if you really need to ensure complete instruction-level control of your robot.

Basic

Basic is one of the most common programming languages for robot hobbyists. This is often used in programming microcontrollers mainly for educational robots.

C++

C++ is a very popular programming language. It provides top=level functionality while you are keeping a good low-level control. A variant of C++ is Processing, which includes simplified codes to make the programming easier.

Java

Java is more developed compared to C++ and offers any safety features to the disadvantage of low-level control. Some producers of microcontrollers such as Parallax are making components for specific use with Java.

Python

Python is one of the most popular languages for scripting. It is easy to learn and could be used to quickly and efficiently integrate several programs.

If you have selected a hobbyist type of microcontroller from a known producer, there's a chance that you can find a book that you can read so you can learn how to program in their preferred programming language. But if you instead prefer a microcontroller from a smaller producer, it is crucial to see what language the controller wants to use and what tools are available.

Conclusion

Thank you again for purchasing this book!

I hope this book was able to help you to learn the basic building blocks of robot building.

The next step is to expand your knowledge in robotics, especially learning advanced programming for your robot.

Finally, if you enjoyed this book, please take the time to share your thoughts and post a review on Amazon. It'd be greatly appreciated!

Thank you and good luck!

Book 2

Human-Computer Interaction

By Solis Tech

The Fundamentals Made Easy!

Human-Computer Interaction: The Fundamentals Made Easy!

Table of Contents

Introduction...34

Chapter 1: Aspects of HCI..35

Chapter 2: The Human Side in the HCI..38

Chapter 3: The Computer Side in the HCI.....................................45

Conclusion..54

Introduction

I want to thank you and congratulate you for purchasing the book, "Human-Computer Interaction: The Fundamentals Made Easy!"

This book contains proven steps and strategies on how to conceptualize and design a computer system that incorporate principles on effective interaction between the user and the device.

Human-computer interaction (HCI) is the study of the interaction between people and computers and the degree at which computers are developed enough to successfully interact with humans. So many institutions particularly academic and corporations now study HCI. Unfortunately, ease-of-use has not been a priority to most computer systems developer. The issue continues to bedevil the HCI community as accusations still abound that computer makers are still indifferent and are not making enough effort to make truly user-friendly products.

On the other hand, the designing task of computer system developers is not simple either as computers are very complex products. It is also true that the demand for use of computers have grown by leaps and bounds outstripping the need for ease-of-use by a significant margin. If you are a computer designer or simply have basic interest in making devices more effective for users, this book will help you a lot.

Thanks again for purchasing this book, I hope you enjoy it!

Chapter 1: Aspects of HCI

Main aspects of HCI

HCI is composed of three main features, namely: the user, the computer, and their interaction or how they work together.

"User" refers to either the individual or the group of users doing things together. An understanding of how the people's sense of sight, hearing, and touch send information is very important. Also, the type of mental models of interactions differs according to the personality of the user. And finally, interactions are also influenced by cultural and national differences.

"Computer," on the other hand, pertains to all technology from desktop to huge computer systems. As an example, if the topic is website design, the computer would then be the website. "Computers" would also include gadgets like mobile phones or even VCRs.

Finally, the "Interaction" is what happens as "User" uses the "Computer" to achieve a certain objective. Humans, of course, are totally different from machines. So the HCI's main intent is to ensure a successful interaction between the two.

In this aspect, adequate knowledge about humans and computers are critical to realize a functioning system. You need to seek inputs from users. Such knowledge would provide much needed information in determining schedule and budget that are crucial to the systems. In effect there are ideal situations and perfect systems. But the key is finding the balance between what is ideal and what is really feasible given the existing situation.

Objectives of HCI

HCI aims to come up with systems that are functional, usable, and safe. Developing computer systems with excellent usability depends on:

- having enough understanding of the aspects that lead people to use technology in certain ways
- being able to devise tools and ways for creating suitable systems
- the development of safe, effective, and efficient interaction

35

- making people the priority

The main philosophy underneath HCI is that the users or the people using the computers always come first. Developers must always be guided by the users' needs and preferences in designing systems. It is the system that should match the requirements of human users and not people changing to suit the nature of the machines.

The primacy of usability

Usability is one of the principal considerations in HCI. It is simply about ensuring that a system can be easily learned and used or be what is called user-friendly. A system is considered usable if it:

- can be learned easily
- can be remembered easily in terms of use
- is effective
- is efficient
- is safe
- is enjoyable

Lack of usability means wasted time, mistakes, and disappointments. Unfortunately, a lot of existing systems and devices have been designed without sufficient attention to usability. These include ATM, the Web, computer, printer, mobile phone, personal organizer, coffee machine, remote control, soft drink machine, ticket machine, photocopier, stereo, watch, video game, library information systems, and calculator.

A good example is the photocopier. If you are not familiar with the symbols on the buttons you will be greatly confused. For instance, the big button with the C on it actually refers to Clear, not Copy. The button used to produce copies is actually on the left side with an unrelated symbol. Devices and gadgets should be easy, effortless, and enjoyable to use.

Analyzing and designing a system based on HCI principles involve a lot of factors that produce really complex analysis because of interactions among many of them. The major factors are:

- The User – motivation, satisfaction, experience, enjoyment, personality. Also cognitive processes and capabilities

- <u>User Interface</u> – navigation, output devices, icons, commands, input devices, graphics, dialogue structures, user support, use of color, multimedia, natural language
- <u>Environmental Factors</u> – health and safety, noise, heating, lighting, ventilation
- <u>Organization Factors</u> – job design, work organization, training, roles, politics
- <u>Task Factors</u> – task allocation, skills, easy, novel, complex, monitoring
- <u>Comfort Factors</u> – seating, layout, equipment
- <u>Constraints</u> – budgets, buildings, cost, equipment, timescales, staff
- <u>Productivity Factors</u> – decrease costs, increase quality, increase innovation, increase output, decrease errors
- <u>System Functionality</u> – software, hardware, application

There are different disciplines representing a wide array of subjects that are covered in HCI. The manifold inputs from these fields have continued to enrich HCI. The disciplines include:

- Cognitive Psychology – limitations, performance predictions, information processing, cooperative working, capabilities
- Ergonomics – display readability, hardware design
- Computer Science – graphics, software design, prototyping tools, technology, User Interface Management Systems (UIMS) and User Interface Development Environments (UIDE)
- Social Psychology – social and organizational structures
- Engineering and Design – engineering principles, graphic designs
- Linguistics – natural language interfaces
- Philosophy, Sociology, and Anthropology – computer supported cooperative work (CSCW)
- Artificial Intelligence – intelligent software

Chapter 2: The Human Side in the HCI

Some of the key aspects that shed light on the human side of HCI are:

1. <u>Perceptual-Motor Interaction</u>. Effective human-computer interface design requires an appreciation of the whole human perceptual-motor system. The information-processing approach is central to the perceptual-motor behavior study and for considering the human factors in HCI. An effective interface design reflects the designer's knowledge of the perceptual such as visual displays, use of sound, and graphics. Also the cognitive exemplified by conceptual models and desktop metaphors as well as motoric constraints like ergonomic keyboards of the human perpetual-motor system.

   Man has gone beyond the use of computer punch cards and command-line interfaces. We now use speech recognition, eye-gaze control, and graphical user interfaces. The importance of various perceptual, cognitive, and motor constraints of the human system is now better recognized in HCI. An effective interface must take into account the perceptual and action expectations of users, the action that is seen with a response location, and the mapping of the perceptual-motor workspaces.

2. <u>Human Information Processing.</u> Aspects of human information processing such as models, theories, and methods are currently well developed. The available knowhow in this field is broadly useful to HCI in general such as in the representation and communication of knowledge and visual display design. An effective HCI requires making the interaction compatible with the human information-processing capabilities. Many things about human information processing have been integrated into cognitive architectures that are now applicable to HCI. These applications include the Model Human Processor, the Act model, the SOAR model, and the Epic model.

   3. <u>Mental Models in Human–Computer Interaction.</u> Studying mental models can help understand HCI by inspecting the processes by which such models impact behavior. For example, mental models of machines can enable both novice and seasoned problem solvers to find new methods for fulfilling a task through more elaborate encoding of remembered methods.

   The Reverse Polish Notation is a great example of this. There is also a general theory that says readers develop a representation in their mind at several levels of what they read. First is the encoding of text, followed by the representation of propositional content of text. Finally, to this text, they integrate world knowledge to form a mental model of the situation described.

Readers also have the ability to look for ideas in multiple texts. They construct a kind of structured mental maps that show which documents contained which ideas even when they did not expect to need it while reading. Mental models are generally considered as semantic knowledge. Focusing on the degree of commonality among team members, for instance, when it comes to knowledge and beliefs, allows quantitative measures of similarity and differences which is the language of computers.

4. Emotion in Human–Computer Interaction. Emotion used to be persona non grata in the field of computer design. It had no place in the efficiency and rationality of computers which were the personification of zero emotion. Recent study findings in the field of psychology and technology show in a totally different light the relationship between humans, computers, and emotions.

Emotion has ceased to be considered only in light of anger generated by inexplicable computer crashes or hyper excitement caused by video games. Nowadays, it is widely accepted that a host of emotions are important part of computer-related activities such as Web search, sending an email, online shopping, and playing computer games. In almost everything now, the emotional systems get engaged according to psychologists.

Studies and discussions on emotion and computers have grown a lot because of dramatic advances in technology. Computers have actually been used to evaluate the relationship between emotion and its correlates. In the same vein, the astounding improvements in quality and speed of signal processing now enable computers to form conclusion on a user's emotional condition. Compared to purely textual interfaces that have very limited range, the multimodal interfaces that can use voices, faces, and bodies are now more capable to a broader range of emotions.

Nowadays, the performance of an interface will be seriously impeded without considering the user's emotional state. Surprisingly, it can earn even descriptions like socially inept, incompetent, and cold. Much remains to be done to successfully incorporate emotion recognition into interfaces. Still, more studies about the interaction between design and testing can help create interfaces that are efficient and effective while providing satisfaction and enjoyment.

5. Cognitive Architecture. A cognitive architecture is a computer simulation program that makes use of human cognition principle based on human experimental data. It also refers to software artifacts developed by computer programmers. Likewise, the term also includes large software systems which are considered hard to develop and maintain.

Right now, cognitive architectures are not widely utilized by HCI practitioners. Nevertheless, it is quite relevant as an engineering field to usability and has important applications in computing systems especially in HCI. It also serves as theoretical science in human computer interaction studies. Finally, cognitive architectures combine artificial intelligence methods and knowledge with data and principles from cognitive psychology.

Presently-known cognitive architectures are undergoing improvements and are being utilized in HCI-related tasks. Two of the most well-known systems, EPIC and ACT, are production systems or built around one. All systems have production rules which differ from architecture to architecture. The difference lies on focus and history although there's a certain similarity in intellectual history. They may have more congruence than differences at some levels either because of mutual borrowings or due to the convergence of the science. The third system, Soar, is a bit different than the first two production system models.

The three production systems, Soar, Epic, and ACT-R were developed to present different types of human cognition but showed more similarities than divergence as they developed. It is not easy to describe a value possessed by architecture as advantage because to others it constitutes a disadvantage. For instance, Soar's learning mechanism is very important for modeling the improvement of users for a period of time. But there are many applications also where Soar's features result to harmful side effects that can cause more difficulty in model construction.

6. Task Loading and Stress in HCI. Stress in the form of task loading is central to HCI. The traditional perspective on stress sees it in light of exposure to some adverse environmental situations such as noise and the focus of attention centers on most affected physiological system. A new way of looking at it, however, stems from the findings that all stress effects are mediated through the brain.

And since the brain is mainly focused on ongoing behavior or current task, stress ceases to be a peripheral issue but that the ongoing task becomes the primary source of stress. And this renders stress concerns that are central to all HCI issues. This means computer-based systems which aim at helping people lessen cognitive workload and task complexity actually impose more burdens and stress on them.

The person's coping mechanism for such stress affects their work performance and personal wellbeing. The environment may vary but some mechanisms for appraising stress in all task demands are the same. So for HCI, certain principles and designs for stress are applicable across multiple domains.

There are several theories of stress and performance and their connection to human-computer interaction. Workload and stress are at times considered as varying perspectives on the same problem. There are some general practices for stress mitigation. But quite important for this topic is setting up effective measures of information processing and mental resources. It also includes expounding on task dimensions that are relevant and their relationship to self-regulatory mechanisms.

It is critical to establish how an individual's appraisal of his/her environment can be influenced by personal traits and states. This is because stress can only be understood vis-a-vis interaction between a person and the environment. Lastly, it is better to treat stress at multiple levels whether physiological or organizational when making practical application. Instead of one-dimensional which is bound to fail, multidimensional is better as it considers the person, task, and the physical, social and organizational environments.

The implication is that HCI researchers and practitioners should go beyond the design of interface displays and controls and focus also on the person aspects. What are the things in the individual that affect performance and the physical-social environment where the human-technology interaction happens? It means that the technical principles at work in that situation are not adequate. They cannot develop a complete description of the relationship between resources and cognitive activities.

7. <u>Motivating, Influencing, and Persuading Users</u>. From its former role as tool for scientists, the spread of computer use to all sectors of society has brought new uses for computers. Among those uses are persuading people to change their attitudes and behavior. Nowadays, it is widely accepted that skills in motivating and persuading people are necessary for developing a successful HCI.

Interaction designers are actually agents of influence which unfortunately they have not yet understood and applied. Yet their works often involve creating something that tries to change people though they may not be conscious of it. Among these works are motivating people to register the software, learn an online application, or have product loyalty. Changing people's attitudes is now a common feature in the success of interactive products.

Depending on the types of product, the persuasion factor can either be small or large. At any rate, anything that needs to be marketed needs to be persuasive. The growing use of computing products and the limitless scalability of software makes interaction designers one of the best potential change agents in the future.

41

Take for example the Web-interaction designers who increasingly are facing more challenge in designing something that will hold the attention and motivation of information seekers. After that, they need to persuade web users to adopt certain behaviors like:
- using a software
- joining a survey
- clicking on the ads
- returning often after bookmarking a site
- buying things online
- releasing personal information
- forming an online community

Being able to persuade people is a measure of success here. But with success comes responsibility. The Web designer needs to make the website credible. The following are some broad guidelines to ensure credibility:

1. Design websites to present the real and practical aspects of the organization.
2. Invest sufficiently in visual design
3. Make websites that people can easily use.
4. Include markers of good quality
5. Use markers of reliability.
6. Avoid too much commercialism on a website
7. Adopt and adjust to the user experience
8. Avoid being amateurish

To sum it all, computer systems have become an inescapable part of everyday life. The interactive experience involving all systems be it mobile phone or desktop can be designed in such a way as to influence the way we think and act. By combining the computing capability with persuasion psychology, computer systems can motivate and persuade. Humans are undoubtedly still superior when it comes to influencing people. But in many areas of endeavor, computer can do what humans cannot even imagine being capable of.

Computers don't sleep and can be designed to keep trying on and on. At the very least, computers provide a new way for modifying how people act and think. Like it or not, the community of HCI professionals is at the forefront of the campaign to make more sensitive and responsive tech products. It can rise to the challenge of helping churn out products that enhance the people's over-all quality of life. Or it can continue being a tool to produce mindless products whose main reason for being into is to make profit for the owners.

8. <u>Human-Error Identification in Human–Computer Interaction</u>. The leap from focusing on human error in technological problems to a less obvious culprit started in the 1940's. It was established during that year that plane pilot error was often designer error. It began to show that design is the key to

substantially reduce human error and this paradigm continued to gather steam particularly in HCI.

It is now common wisdom that human error can be as often as the product of a defective design or as a person making a mistake. The inadequate design fosters activities that lead to errors. A groundbreaking outcome of this new philosophy is that errors are now viewed as totally predictable events instead of seeing it as unpredictable occurrences. This makes errors avoidable.

So errors became instances where planned series of steps and activities fail to realize intended results independent of any outside change agencies. If errors are no longer random, then it can be identified and predicted ahead of time. What partially drove this line of thinking are the accidents that happened in the nuclear industry that is hungry for preemptive solutions. This has led to the formulation of several human-error identification (HEI) techniques.

Although evaluative and summative in nature, these HEI techniques that employ ergonomics methods can now be used in formative design stages especially in analytic prototyping. For instance, the entry of computer-aided design such as in architecture has profound impacts on prototyping. It made possible what was considered as impossible or too prohibitively costly design alteration at the structural prototyping stage.

The three main forms of prototyping human interfaces have been identified namely: functional analysis, scenario analysis, and structural analysis.

Functional analysis includes consideration of the functional range supported by the device. In comparison, scenario analysis is exemplified by consideration of the device in relation to events sequence. An example of the structural analysis, on the other hand, is the use of user-centered viewpoint in a non-destructive testing of the interface.

One compelling example of the crucial role of design in predicting and minimizing errors concern human error identification (HEI) tools like the TAFEI or Task Analysis for Error Identification. The results of the application of TAFEI on interface project designs show how it can improve systems and its relevance to other ergonomic methods. It served to validate what has been long suspected when it comes to error-design relationship as follow:

- Structured systems like TAFEI results to reliable and trustworthy error data;

- Most errors resulting from technology are totally predictable;

- To improve design and reduce errors, ergonomics methods should be employed in formative design process.

Exploring design weaknesses through tools like TAFEI will go a long way in developing and producing devices and gadgets that are tolerant to error.

Chapter 3: The Computer Side in the HCI

The salient points when it comes to the computer side in HCI include:

1. Input Technologies and Techniques. Input devices which are also a classification of computer can detect physical aspects of places, things, and, of course, people. However, its function is never complete without considering the visual feedback corresponding to the input. It is like using a writing instrument without something to write on. Input and output should always go together.

   And in devices with small screens, this is only possible with the help of integrated sensors. If the user or human characteristics are important in a maximized HCI environment, so are input technologies with enough sophistication to meet user-machine interaction requirements. Users can only achieve the task objectives by combining the right feedback with inputs. In this regard, the HCI designer should take into account the following:

a. the industrial and ergonomic design of the gadget

b. the physical censor

c. the relationship among all interaction techniques

Input gadgets have many properties that apply to the usual pointing devices or mobile items with touch input. These pointing devices include the: mice, trackballs, isometric joysticks, isotonic joysticks, indirect tablets, touchpads, touchscreens, and pen-operated devices. The mice or mouse, of course, is one of the most popular as anyone who has ever used a computer knows. Because of its inherent advantages for individual users where it can easily be used by most people, it is one of the most preferred pointing devices.

Touchpads are most well-known to laptop users. These are small tablets that are sensitive to touch and which are usually featured on laptop units. Touchscreens on the other hand are tablets that are sensitive to touch which are placed on a display. It is increasingly becoming the tool to beat because of the proliferation of smart phones and other hand-held devices.

There are input models and theories that are quite helpful in evaluating the efficacy of interaction strategies. But it would be most beneficial to readers here to focus on current and future trends for this feature. Interactive system designers should go beyond the usual things like graphical user interface and pointing ideas when it comes to inputs.

They must delve deeper into more effective search strategies, sensor inputs for new data types, and techniques of synthesis to make much better sense of data. Better search tools will enhance navigation and manual search regarding file systems. One outstanding development is the breakthrough in the development of more advanced sensor inputs such as technologies for tagging and location.

It allows computers to identify physical objects and locations that have been tagged, and to detect their location and distance to other devices through signal strength analysis. These sensors are making interface personalization much easier. This development in interaction also has great implications for data mining and techniques for machine learning. Continuous improvement in structure synthesis and extraction techniques is invaluable in this data-rich era.

An overriding aim in HCI is to achieve dramatic advancement in humanity's interaction with technology. The computer side of this presents limitless possibilities but the cognitive skills and senses of man will be relatively stagnant. Our holding, touching, and object-movement are not the result of technology-like progress but a product of our human limitations.

2. Recognition- and Sensor-Based Input for Interaction. Computers are able to manipulate physical signals that have been transformed by sensors into electrical signals. Sensors have found their way into various fields of industry such as robotics, automotive, and aerospace. It has also found vast applications in consumer products.

The computer mouse is a very good example. Imagine that simple-looking device equipped with algorithms that process images and specialized camera that enables it to be unbelievably sensitive to motions. It detects movement at the rate of a thousand of an inch several thousand times per second. Another interesting device is the accelerometer that detects acceleration due to movement and continuous acceleration because of gravity.

Digital cameras now make use of accelerometers to save a photo. Laptops also are equipped with accelerometers for self-protection. When the laptop is accidentally dropped the accelerometer enables the hard disk to secure the hard drive prior to impact. With smartphones, the goal is for motion sensing for the purpose of interaction such as determining the walking pattern of users. Generally, HCI research on sensors dwells on its usage to improve interaction.

Sensor studies are either to broaden input options or build new computing forms. The new forms include mobile devices that recognize locations and places that are sensitive to the presence and needs of its inhabitants. Still there are far more advanced goals and applications like in robotics.

There is a race to develop machines that will behave and think like humans or at least complement their capabilities. It has many critical applications such as in nuclear power accident mitigation. One worry, of course, is that it will end up in the military. But in safety, mobile computing, entertainment, productivity, affective computing, and surveillance, sensors are finding widespread application.

An intriguing side note here is the idea of developing a sensor to enable computers to detect and accordingly react to the frustration of its user. The computer's response could be something like playing relaxing music. Sensing could be in the form of the user banging on the keyboard in frustration. A microphone could react to the yelling of the speaker or the webcam could sense scowling.

In general, the potential of interactive sensing is quite good. The degree of progress across the whole computing spectrum actually gives the impression that sky is the limit. Advances in nanotechnology, CPU power, and storage capacities will continue to produce more outstanding innovations in the computer side of HCI.

But what is driving the unprecedented growth of the sensor-based interactive systems is the dizzying expansion in devices outside the old desktop computer. It is hard to keep track of the explosive proliferation of smart phones, tablet PCs, portable gaming devices, music and movie players, living room-centric computers, and personal digital assistants. Computing is becoming part and parcel of our daily life and our environment.

Through recognition techniques and sensing systems, task-specific computing devices will be developed instead of general functions. It will also pave the way for different types of interaction style in HCI. This activity-specific interactive systems development will further hasten innovations on a much broader array of practical applications.

3. Visual Displays. Timekeeping has always been one function that man has strived for a good visual display. Today's smartwatches which are actually wrist computers sport stunning visual displays. It is no longer limited to displaying time but is multifunctional. Some brands can pinpoint exact location in the planet through a global positioning system. Others can show heart rate while a number can be personal digital assistants.

The main idea behind wearable computing is that the human body is wearing the visual displays. One major way people use wearables is to put the display on one's head making user's hands free to work. It is called headmounted displays or HMDs. The screen-based is one category of HMDs. It makes use of the retinal-projection method which projects images on the retina of the eye.

An alternative method is the scanning displays which scan images onto the retina pixel by pixel. A second type of display which is actually much bigger in scope and the most widespread is the hand-held and wrist-worn displays. They are in mobile phones, media players, wristwatches, and other portable gadgets. Apple is one of the global leaders in this field and the most well-known. Even textiles for clothing are now being used for such technology in what is now known as photonic textiles-fabric.

Multicolored lighting systems were merged with the fabrics for its electronic information function without affecting the cloth's softness. It has sensors, GSM, and Bluetooth! Photonic fabrics have great promising applications in the areas of personal health care and communication.

4. <u>Haptic Interfaces</u>. Haptic interface refers to a device for sending feedback that produces sensation of weight, touch, rigidity, and other aspects through the skin and muscle. This force feedback mechanism is designed to enhance computer-human interaction. Because haptics are done through actual physical contact, they are not easy to synthesize unlike the sense of sight and sound that are gathered through the eyes and ears.

   The genius in the haptic interface is that it simply makes use of the body's own highly sophisticated receptor system. The haptic feedback is made possible through synthetic stimulation in the skin and proprioception.

   Proprioception involves something deeper than the skin – the muscle and skeleton. The mechanoreceptors in the body enable its detection of contact forces received from the environment. Body receptors sense velocity, skin stretching, vibration, and edges of objects. Haptic interfaces are more widely applied in the field of virtual reality than in information media and related devices are now available commercially.

Two of the most important research needs on haptic interfaces in the future concern the psychology in haptics and safety considerations. Safety is a crucial issue as insufficient actuator control can lead to injuries for users. Control problems may occur with the tool displays and exoskeleton. Unintended forces or vibration may pose danger to the user.

A locomotion interface that holds a user's body can cause serious physical damage if control is inadequate. It requires proven safety equipment that amply protects the walker and this should be a major objective of research. A much safer alternative is a system where the user does not wear any equipment during the interaction.

The psychology in haptics on the other hand requires more studies on muscle sensation as most existing findings are on skin sensation. Among the few promising findings relate to Laderman and Klatzky's work (1987) on force display and their recent study of forces distributed according to space. Their psychological findings have very promising applications in the development of haptic interface. A lot of obstacles need to be overcome before usage of haptic interface becomes widespread.

Though men cannot do without haptics in real life interaction, it is still of limited use in HCI. One can say haptic interface is still in infancy with its 10-year background. So eventually its time will come just like image displays (e.g. TV and movies) which started 100 years ago. For now, what are available are a few haptic interfaces with limited functionality and high cost. At the very least, haptic interface is a new very promising frontier in HCI with immense potential contribution to man's quality of life.

5. Non-speech Auditory Output. Sound is one of the key aspects that complete our interaction with our environment. But where speech is direct and necessitates focus and attention, non-speech sound is more diffused and provides a different class of information.

   Non-speech sounds include sounds from the environment, music, and sound effects. Nevertheless, speech and non-speech sounds complement each other just like text is complemented by visual symbols. Non-speech sounds can give information in a shorter period of time than speech.

   Right now, the non-speech field needs more research. The user interface is a much more effective tool for HCI when it employs a combined visual and sound feedback. This sound-visual combination has complementary function as well. Visual gives specific information about a small area reached by our eyes but sound or the auditory system provides more general information from beyond our focus.

   Our senses are the key to our effective interaction with the external world. These senses in turn bring more dimensions in information as they enhance one another. These principles are very useful in a multimodal HCI by adding non-speech sound output to the graphical displays. An example of this application is focusing our eyes on one task like editing a manuscript while monitoring other aspects in the machine through sound.

Reliance on visual sense which is more prevalent at present can be problematic. One problem is there could be visual overload which means the user could miss lots of information. Or simply that the viewer cannot look at everything at the same time at all times. Sound can help eliminate that situation by giving information to the user that the eyes could not see.

This interdependence between visual and audio could make information presentation far more efficient. Non-speech sound is mainly used in games' sound effects, music, and other multimedia usages. It is commonly employed in creating a certain mood for the item like in movies. In HCI, sound is used to provide information particularly those things that a user does not see or notice such as what is going on in their computer systems.

It is useful to use non-speech sound in HCI for many reasons. Seeing and hearing in the human body is first of all interdependent. The eyes can give information that is high-resolution only in a limited area of focus. But sounds can be received from all sides of the user: front, above, below, and behind. This not only provides direct information but also tells the eyes where to look to get more useful data. In fact, at times reaction to sound stimuli is faster than what is seen.

Non-speech sound can therefore help in reducing large display overload which can cause users to miss important data. This is especially true in large graphical interfaces that use multiple monitors. Using sound to present some information would reduce screen space. It would also lessen the volume of information that should be on the screen. This is most relevant to gadgets with small visual displays like smartphones and PDAs.

Non-speech sound would also decrease demands on our visual attention. For instance, a user who is walking would miss much information as he looks at his device's visual display because of competing attention from the traffic or uneven surface where he is walking. In fact, if the information is in sound, he does not have to look at his device at all.

Our sense of sound is also underutilized. Yet as exemplified by classical music, its intricate organization can make, say a symphony, a powerful tool for transmitting complex information. The beauty with sound is that it grabs attention. It is easy to avoid looking at something but hard to ignore sound which makes it very effective in sending important information. Likewise, certain things in the interface look more natural in sound than in sight.

Finally, non-speech sound will allow visually-impaired users to use computers. Newer graphical displays have, in particular, made it even harder for them to operate the device. Research has been extensive in the HCI application of non-speech sound in a wide range of topics.

There are two main areas of growth where the application of non-speech sound has the best potential. One is in the creation of multimodal displays that utilizes all available senses. This means integrating sound with other things like force-feedback and tactile apart from sight. The other area is in wearable and mobile computing gadgets that also use multimodal displays.

As mentioned, the screens of these devices are small and sound will reduce the need for screen space.

6. <u>Network-Based Interaction</u>. Networked interfaces have modified our perception of society and the world at large particularly with the Web and now mobile devices. There are several roles that networks play in HCI. The first is as an Enabler which refers to things that can be done only with network. The second is as Mediator which pertains to problems and issues caused by networks.

Third is as Subject which focuses on managing and understanding networks and fourth as Platform which dwells on interface architectures and algorithms. Network includes both the wire-based and the wireless world. Things are rapidly changing especially in the wireless networks. These changes can be classified in two dimensions, namely:

- **Global vs. Local** – refers to the distance by space between the connected points such as machines in the office to global networks like the Internet.

- **Fixed vs. Flexible** – pertains to the nature of the links between points such as fixed devices and gadget that configures itself. More changes are coming because of spreading wireless links. One example is being able to gain access to internet connections and printers of another office by simply plugging a portable device into the Ethernet.

Traditionally, LANs belong to local-fixed category while Internet is global-fixed. Hand-helds like cell phones are also categorized as global-fixed because phones are fixed and independent of location. The internet makes use of domain names which are fixed like URLs. Some phone technologies like GSM and GPRS are classified as global-fixed because it is possible to send content that is based on location. Also the enlarging data capability is enabling services to handle huge media content.

What set these technologies apart, however, are the connectivity model and the charging which are usually by data use or fixed charge. There are a number of current and new technologies from the local-flexible type. These include the Wi-Fi, infrared, Bluetooth, and ZigBee which permit flexible connections among personal gadgets. With them, a computer device can utilize a mobile-phone modem or a headset with Bluetooth can make connections with a phone, wireless. Unfortunately, these capabilities also enhance unsavory activities like illegal equipment accessing, hacking, and surveillance.

7. <u>Wearable Computers</u>. Computers have become like appendage to many office workers. But it is hard for those using mobile devices to get the information

they need. In a mobile situation, existing interfaces will hamper the user's main task. Users will be forced to prioritize the device instead of the environment. The need is for a wearables design that helps fulfill not obstruct the task.

A framework that can be very useful in creating good designs of computer interfaces which are wearable is CAMP. This framework addresses different factors that may impinge on the effectiveness of the design such as body closeness and how it is used. CAMP stands for:

- <u>Corporal</u> – which means absence of discomfort to users during physical interface with the wearable.
- <u>Attention</u> – interface design should allow user to focus both on the real world and virtual reality.
- <u>Manipulation</u> – there are adequate controls which are easy to manipulate particularly in a mobile environment.
- <u>Perception</u> – Design must enable user to quickly perceive displays even when mobile. So displays should be easy to navigate and simple.

Outside offices and buildings, an attractive option for a user to have access to a computer interface is through wearables. There are challenges however that need to be addressed to fulfill the tasks in terms of contextual awareness, interface, adaptation to tasks, and cognitive model. These include:

- **Modalities of Input/output** – the ease of use and accuracy of modalities developed that try to copy the human brain's input/output capacity are not yet satisfactory. Frustrations bedevil users when there are inaccuracies. Also the computing requirement of these modalities is way beyond what low-weight wearable devices have. Input devices which are simple to use are needed.

- **Models of User interface** – there is a need for extensive experimentation in using applications involving end-users.

- **Capability-applications matching** – evaluation and design of interface should prioritize development of most effective way to access information and avoid creating additional features.

- **Simple methodology in interface evaluation** – current evaluation approaches are too complicated and time-consuming making them unsuitable in interface design. What is needed is an evaluation methodology that addresses frustration and human errors.

- **Context awareness** – for context aware computing to be realized, several questions must be answered. These include application models that integrate the social and cognitive aspects, social and cognitive mapping of inputs from many sensors, anticipating the needs of users, and interacting with the users.

Conclusion

Thank you again for purchasing this book!

I hope this book was able to help you to gain useful knowledge and understanding about human-computer interaction.

The next step is to apply what you have learned.

Finally, if you enjoyed this book, please take the time to share your thoughts and post a review on Amazon. It'd be greatly appreciated!

Thank you and good luck!

Book 3
Raspberry Pi 2
By Solis Tech

Raspberry Pi 2 Programming Made Easy!

Raspberry Pi 2: Raspberry Pi 2 Programming Made Easy!

Table of Contents

Introduction...58

Chapter 1: Raspberry Pi – The Basics...59

Chapter 2: Hardware Accessories...61

Chapter 3: Installing the Operating System In Your SD Card.........64

Chapter 4: Booting OpenELEC & Raspbian for the First Time On Your Pi..69

Chapter 5: How to Get Around in the Raspberry Pi's Unix-based Command Line Interface...71

Chapter 6: Python Programming in Raspberry Pi...........................77

Conclusion...80

Introduction

Greetings. Thank you for purchasing this book Raspberry Pi 2: Raspberry Pi 2 Programming Made Easy. If you're reading this book then most likely you are here to learn about his nifty new little computer called the Raspberry Pi. In this book, we'll teach you everything that you need to know to get up and running with this little, but powerful computer.

We'll teach you about its parts, its specifications, setting up the operating system, its capabilities and limitations, and ultimately how to do basic programming. We'll discuss the best programming language that works best with the Raspberry Pi and create your first program with it. Without further ado, let's start your Raspberry Pi journey.

Chapter 1: Raspberry Pi – The Basics

What is a Raspberry Pi?

So what is it exactly? We'll it's a small credit-card sized single board computer that is intended to help people learn more about programming, how the computer works, etc. The CPU or Central Processing Unit is basically a system-on-a-chip. What that means is it pairs an ARM processor that is used by a lot of embedded systems and cellular phones with a Broadcom GPU (Graphics Processing Unit), which is a fairly powerful graphics processor that's capable of displaying full resolution 1080p HD video.

The amount of RAM (Random Access Memory) that the Raspberry Pi 2 has is 1 Gigabyte. Its previous iteration, the Raspberry Pi 1 Model A+ and Raspberry Pi 1 Model B has 256MB and 512MB RAM respectively. The RAM is shared by both the CPU and the GPU.

The Raspberry Pi 2 starts at $39.00 and that includes just the board. On the board there are 4 USB ports, 40 General Purpose Input/Output (GPIO) pins, a full 1080p HDMI port, a Network/Ethernet port, a 3.5mm audio and composite video jack, a Raspberry Pi (CSI) Camera interface, a Video (DSI) Display interface, a Micro SD card slot, and a VideoCore 3D graphics core. The GPIO is basically for more advanced users who are going to be adding Arduino accessory boards, ribbon cables for communication with other hardware devices, and major electronics projects; robotics, sensors, etc.

Why Is it Cool to Have a Raspberry Pi?

Most people think the Raspberry Pi is cool because it's a fairly complete computer, it runs on very little power, it's small, and it will not burn a hole through your wallet; it's $39.00, which is less than the cost that you would pay for a dinner with a friend or a loved one. The Raspberry Pi also helps people that are new to computer hardware get into it and get their hands dirty without the cost and risk associated with more expensive standard hardware.

Another thing that's cool about it is that the Graphics Processing Unit is pretty powerful. It can play 1080p HD video, which makes it really attractive to a lot of people as a Media Center PC. In fact, most people buy the Raspberry Pi 2 for that specific reason because it is a cheap multimedia PC.

What Can You Do With a Raspberry Pi?

So what can you do with this thing? Well, just like what was previously mentioned, if you're really new to computer hardware, the Pi can help you learn about those individual components of most modern computers. And if you want to learn how to do programming, the Raspbian operating system that comes with the Raspberry Pi 2 comes with a lot of tools for programming to help you get started.

So you might be asking, why wouldn't you start learning how to program with the computer that you already have? Well, you can. And that's perfectly fine. However, you can think of the Raspberry Pi as your little playground and your test computer that you don't have to worry about breaking if you screw something up. Even if you do break it, you'll only be losing $39.00.

But if you don't care about all that and you don't care about programming or about learning the hardware and so forth, the Raspberry Pi 2 also makes a really great little Media PC. You can get a small case for it, hook it up to the network, install another operating system that's based on Linux called OpenELEC, which is basically a back-end for XBMC (Xbox Media Center), plug it into your TV, and you're ready to start streaming video and audio content on your high-definition TV.

Chapter 2: Hardware Accessories

In this chapter, we're going to talk about some of the hardware and accessories that you're going to need to get started with the Raspberry Pi 2. First, let's start out with the bare necessities. The two things that you'll definitely need are power and an SD card.

Power Requirement

Let's take a moment here and talk about power because it's important. The Raspberry Pi is a little picky about its power source. Learning about what you need here for power could save you a lot of headache later down the road. The Raspberry Pi needs 5 volts. That should be anywhere between 4.75 volts minimum and a maximum of 5.25 volts. It should be at least 700 mAh, but the recommended is 1000 mAh.

The reason for the 1000 mAh recommended power is because when you start plugging in a keyboard, mouse, network cable, and other peripherals, these all pull power away from the system. This is also why some Raspberry Pi 2 users say that 1500 mAh to 2000 mAh is even better. Of course, if you're only doing projects close to your desktop computer and you don't really plan to run the Raspberry Pi 24/7, you could just run it off of your desktop system's power with just a USB cable if you want.

SD Card Storage

There is not on-board storage on the Raspberry Pi. So you need an SD card to run the operating system from and also to store any files. Get at least the Class 4 SD storage. It is recommended that you get the Class 6 or Class 10 if you don't mind spending a little bit more.

However, if it's not available, just get the best one that you can. The cost difference honestly isn't that much, and SD cards are getting cheaper all the time. You'll also need a card reader to transfer the operating system image that you're going to download off of Raspberry Pi's website from your desktop computer to the SD card. You can also buy SD cards with the operating system pre-installed from whosoever your Raspberry Pi vendor is if ever you want to skip the operating system installation on the Pi.

If you want to try out multiple operating systems, you can do that with a single card. However, just keep in mind that you'll overwrite any of your settings that you've played around with in the operating system.

Interacting With the Raspberry Pi

You'll need to decide on how you want to interact with your Raspberry Pi. You can either do it headless over a network, with a keyboard, mouse and monitor, or you can do a combination of both. For network access, you're going to need a network cable; either a CAT5e or CAT6 to connect to your router. There's also a USB WiFi option, which allows you to connect to your router wirelessly.

If you're going headless, you don't really need a USB keyboard and mouse. They are both optional. It just depends on what you want to use the Raspberry Pi for. Just like what was previously mentioned, if you don't want to use the desktop interface, you can skip using a mouse and just plug in a keyboard. If you're using it as a media center and navigate XBMC, a wireless Bluetooth keyboard is a fantastic option if you want to sit on your couch and navigate your media center.

If you're only planning to login remotely using SSH, you can skip the keyboard entirely, or if you want you can just borrow a keyboard and mouse from another machine for the initial setup before you actually go headless over the network. Raspberry Pi should automatically get an IP address from your router. After your router gives an IP address to your Pi, you can just log in to your router to see the exact IP address that was assigned. You can then connect via SSH using that IP address.

If you don't want to find out what the exact IP address is via the router and instead want to find it out locally, you can pull up the command terminal in Raspbian OS and execute an "ifconfig" command. You can then unplugged your peripherals and go headless from there.

Displays

Moving on to displays, if your TV or monitor has an HDMI input, that's great because all you'll need to get is a cheap HDMI cable. If your monitor only has DVI-in, you'll need an HDMI to DVI cable or an adapter. Just keep in mind that if you do that, the audio signal that's coming out of the HDMI port will be ignored when you're using a DVI converter.

Depending on your setup, you can use external speakers or a splitter. If you're going to go headless, you can skip purchasing a separate monitor specifically for your Pi and just borrow one for the initial setup if you want to. Otherwise, you can still do most of the setup across the network if you are connected via SSH.

Raspberry Pi Casing

Moving forward, even though it may look cool keeping your bare board out with a bunch of wires just hanging off, it is probably not the safest thing to do for your Raspberry Pi. So it is recommended that you get a case for your Pi. You can get a prefab case for as little as $5.00. It is completely optional but it does help protect the board. A number of people have made some pretty cool custom cases, all of which are available for purchase on-line.

Chapter 3: Installing the Operating System In Your SD Card

In this chapter we're going to be looking at operating system options and how to actually install it in your Raspberry Pi. First off, you're going to need to consider what your goal is for your Raspberry Pi. Is this going to be a learning or tinkering board? Are you going to be programming or do you just want a quick and cheap home theater PC? For a more general purpose operating system or for the tinkerer and programmer, most people recommend going for Raspbian OS.

The latest version of Raspbian at the time of this writing is Raspbian Jessie Kernel Version 4.1, which was released in November 2015. For a home theater PC, most people recommend getting either OpenELEC or RaspBMC. If you're just not sure and you want to try multiple operating system versions, get two or three SD cards and you can just install a different operating system on each of them. You can swap them out depending on what you feel like doing with your Raspberry Pi. Below is the link where you can download the various operating systems for your Pi:

https://www.raspberrypi.org/downloads/

The Operating system installation method that we're about to discuss is mainly for those who have a Mac or a Linux desktop. If you have Windows, the concept is still pretty much the same. You'll just use a different program to get the disk image onto your SD card with Windows.

Step 1

Regardless of which type of PC you're using, you're going to need to download the disk image or images of your preferred operating system for your Raspberry Pi. These are by no means the only options, just the ones that the Raspberry Pi community recommends in order to get you started. The first one, as we've mentioned before, is Raspbian.

There's also another version that came out called Occidentalis, which is actually based on Raspbian but has a more educational slant to it and includes a lot more educational tools. For a home theater PC, you'll want to install either OpenELEC or RaspBMC. Both of these are built around the XBMC (Xbox Media Center).

Step 2

Once you have chosen your operating system, click the download link and choose a location in your computer that you're going to easily remember. Once your download is complete, it would probably be in a .zip file format so you'll need to unzip it and you should now have an image file with an .img file extension. There are a lot of different methods to get this image onto the SD card, but if you're on Linux or Mac, we're just going to use the "dd" command to write the image on the card.

We're going with this method mainly because it's installed in almost every UNIX-based system and it should work pretty much the same on all of them. Before we proceed in talking about DD in Mac or Linux, if you're a Windows PC user, you're going to need to download the win32 disc imager. Once you've installed it, you need to run it and tell it where the operating system image file is located on your hard drive and what drive letter to write the image to for your SD card.

When choosing the actual drive letter that points to your SD card, make sure that the SD card you're using is blank since everything inside it is going to be overwritten. Getting back to Linux and Mac, you want to open up a new shell or a terminal as it's called on the Mac, and then navigate to the folder where you downloaded your disk image. In most cases it would be in /Downloads/Raspberry Pi/Raspbian.

Again, keep in mind that you will be overwriting everything on the SD card that you're going to be using for this installation. So use a fresh card or one you don't mind overwriting. At this point, make sure you do not have your SD card reader plugged in yet. The reason for this is because we need to find out which device our SD card will be using.

In your terminal or shell, type "df" (without the quotes) and then press ENTER. Now you should see a list of devices and the folders where they are mounted. This should be any of your internal hard drives or external hard drives and so forth. Now is the time when you want to plug in your SD card into your reader and then plug the reader into an open USB port.

Just wait a few moments for the card to mount. This can take a few seconds. After a few seconds, use the "df" command again. You should now see something different; an item that's listed as either no-name or volume/untitled. On the side you should see something listed like dev/disk3s1 or on Linux it would be something similar to dev/sdb1 or sdc1 or something like that next to it.

Drive assignments may vary from user to user so just take which device is listed next to that volume carefully. You want to make sure that you're using the SD device and not one of your hard drives. DD is an unforgiving command and it will overwrite without prejudice and without asking you if you're sure about doing so. So just be very careful. We can't stress this enough. You don't want to overwrite your main hard drive.

Once you know the device that you're going to write to, remember or write it down. Like what we mentioned earlier, it should be something like dev/disk3s1 on a Mac or something similar to dev/sdc1 on Linux. We're going to change this a little bit since we actually want to write the disk image to the entire card not just a partition on the card, which is what we have mounted right now.

So what we need to do now is unmount the SD card's partition before we can write to the entire card. We don't want to eject the card, we just want to unmount it. On your Mac you can open up your Disk Utility. It's in the same folder as your terminal. You then have to select the volume under the SD card device and click "unmount." You can also do this from the terminal, but you'll need to be on the superuser account. So to use Disk Utility using a superuser or root account, you can type the command below:

```
$ sudo diskutil unmount/dev/disk3s1
```

On Linux you would do a similar thing. You would type in the command below:

```
$ sudo unmount /dev/sdc1
```

You can type the command above or whatever your device assignment happens to be. One thing to remember is that the sudo command will always ask you for the administrator password in order to be able to execute the command, regardless of whether you're using a Mac or a Linux system. Once the SD card's partition is unmounted, we're going to run DD with 3 arguments; "bs," "if," and "of." Those stand for block size, in file, and out file respectively. You'll run the command exactly as below:

$ dd bs=1m if=2015-12-24-jessy-raspbian.img of=/dev/sdc1

Looking at the command above, you're going to be putting a value for block size; specifically 1m which stands for 1 megabyte, for "if" you'll be putting the name of the image file, and "of" is the location where you stored the image file. Regarding the location of the image file, we're going to change that a little bit. For example, if your device is listed as /dev/disc3s1, you want to change the "of" argument to /dev/rdisk3s1.

This may be different depending on your system so changing this will make sure that you use the entire SD card and not just the partition. In Linux for example, the "of" value would probably be something like /dev/sdb1 or /dev/sdc1. In this case, the letter after the letters "sd" could be different depending on your system. So in order to write to the whole card, you just need to take away the number "1." You should have something similar to the one below:

$ dd bs=1m if=2015-12-24-jessy-raspbian.img of=/dev/sdb

So that's pretty much of it. Once you're confident that you have the correct arguments for the DD command, hit ENTER. You're going to wait a few minutes while the SD card is being written on. At this point, you won't see any output while this process is ongoing. It will probably take a few minutes so just be patient.

If you're on Linux and you get an error, you might need to change the "bs" value from 1m to 1M (capital M). For some reason the case of the letter matters to a Linux system. Once this has been completed, you should be back to a command prompt. And assuming everything worked properly, you will now have a bootable operating system on your SD card.

You are now ready to boot your Raspberry Pi for the first time and start configuring your system, which we will cover on the next chapter.

Chapter 4: Booting OpenELEC & Raspbian for the First Time On Your Pi

In the last chapter, we installed the operating system on the SD card. Let's now go ahead and insert the SD card into the slot on the underside of the Raspberry Pi. Keep in mind that the metal contacts of the SD card should be facing the underside of the board. First, we'll need at least a keyboard, a mouse, and a monitor. Go ahead and plug the keyboard and mouse to any of the available USB ports. Second, go ahead and plug in the Ethernet CAT5 or CAT6 cable on the Ethernet port on the Pi. Make sure that the other end of the Ethernet cable is plugged into your router. Third, go ahead and plug in the HDMI cable for your monitor.

Do not forget to use an HDMI to DVI converter if you have a DVI monitor. Lastly, plug in the microUSB cable of the 5V power supply to the microUSB port on your Raspberry Pi. Once you have everything plugged in, you should see the LEDs on the board start to light up. You will also see that the boot cycle has now started on the monitor. Just wait for login prompt to appear once the boot process has finished.

Logging In

Once you have the login prompt, you want to use the default username which is "pi", and the default password, which is "raspberry" in all lower case. Once you type those in and press ENTER, you're in.

To start the desktop interface just type "start x" and press ENTER. Most people should be pretty familiar with how this desktop works. It's very similar to most other modern GUI interfaces. Let's now take a look at some of the applications that are available to you once you've successfully started the desktop interface of your Raspberry Pi.

- Scratch – This is basically a programming teaching tool. It is very similar to Karel. If you've ever taken a programming class, they use Karel the robot to teach you basic programming concepts.

- Python – Python is basically a very versatile and robust programming

69

language. It is used by many programmers and companies to make web based applications.

- Midori – Midori is a lightweight web browser. It takes its roots from Mozilla Firefox.

- File Manager – The file manager is very similar to the Window file manager. It lets you see the different directories and storage media that's associated with your Raspberry Pi. It also lets you see the important files in your Raspberry Pi.

- Terminal – Since Raspbian is basically a derivative of Debian Linux, it's no surprise that it also has a terminal similar to Linux. It has the same functions as a Linux terminal. It can also run all the terminal commands that you can run in a Linux system.

Now that we've pretty much seen what the Raspbian operating system looks like, let's go on ahead and see what OpenELEC looks like in a Raspberry Pi. Go ahead and mount the image of OpenELEC onto your SD card. The mounting process for OpenELEC is pretty much the same as Raspbian. If you're still logged into Raspbian on your Raspberry Pi, you must log out and shut the Pi down.

You shut the Raspberry Pi down by using the "halt" command. Again, in order for this command to work, you have to be a superuser. So you need to type in the word "sudo," then put a space by pressing the spacebar, type in the word "halt," and then press ENTER.

After you press ENTER, the Raspberry Pi will initiate the shutdown process. Once finished, you can now switch to the SD card with the OpenELEC operating system. Once the card is inserted, you can start booting your Raspberry Pi with the OpenELEC operating system. As you can see, you can pretty quickly change out cards and be up and running with a different operating system pretty quickly.

Once you have OpenELEC up and running, go ahead and explore the different parts of the operating system, see the different programs it has, and find out what it can do. This is the perfect time to explore and learn what the Raspberry Pi and its operating system is all about.

Chapter 5: How to Get Around in the Raspberry Pi's Unix-based Command Line Interface

In this chapter, we will be going through some of the more basic commands that you're going to need to know in the command line shell for getting around the Unix-based file system. If you're running Raspbian, you'll need to know some of the standard Unix commands. Like most systems in Linux or Unix, files are stored in directory hierarchies.

You can also call them folders if you want to feel more comfortable with that term. When you login, unless you have a custom setting, you will most likely begin in your home/username directory. For Raspbian, the default username that we've been logging in with is Pi. So your home directory is going to be /home/Pi.

Most users on the system, except for root, will have their home directory stored within /home. The main exception is the root user's folder, which is stored in /root. On your command prompt you will see your home directory represented as a tilde symbol "~", and you can use that to easily get back to your home folder anytime.

Moving forward, let's get into working with directories. The first thing you might want to know is where you are and what directory you are located in. The first command that we're going to start with is PWD. PWD stands for Print Working Directory, which tells you the current directory that you're located in at that moment. You'll find that most commands in Unix are abbreviations of the activity that they perform.

As you can see from the output after you typed in PWD and pressed ENTER, you're in /home/pi. If you want to change to another directory, you need to use the CD command. For example, if you want to go to the top level directory—also called the root, which is not to be confused with the root user home directory— you type in this command:

    $ cd /

cd / represents the main directory; and the slash that is used is a forward slash. Many people mistakenly use the term backslash when they really mean forward slash. We call the "/" a forward slash because we read from left to right. So a slash leaning to the right is in the forward direction.

Typing the aforementioned command and pressing ENTER will bring you to the top level directory. There is nothing above this top level directory. It only has subdirectories beneath it. With that being said, let's now take a moment to look at a few subdirectories beneath the main root directory.

The first one that we're going to look at is "etc." Etc stores many of the system's configuration files so get familiar with this folder. If you want to change configuration settings, you're going to come here often. To go to the etc subdirectory from the root directory, type in the command below:

$ cd /etc

Now let's take a look at what's inside etc. We do that using the "ls" command. Ls is short for list and it just lists directory contents. ls also have some options associated with it that allows the user to change the way the listing looks.

Let's go ahead and take a look at a few common options of ls:

- -a – lists all files including hidden files in the directory. Files beginning with a dot "." are considered hidden. So anything that you name with a dot preceding the file name will be hidden automatically, unless you use the -a option to list the files.

- -l – lists all the files lengthwise down the screen.

- -p – puts a forward slash after the directory, which is inside the current directory that you are viewing. This lets you easily distinguish directories from files.

- --color – this will give you a colorful directory listing. However, this

72

doesn't always work. This command works depending on the system and if you're logging in remotely.

You can also combine multiple options with a single dash. Look at the example below:

$ ls -lap

Typing the command above will give you all the options that you've indicated together with the list command. Now, to go back a directory, you have two options. In this particular case we can either type cd / to take us back to the main root directory since that's where we started before going into etc, or we can type the command below:

$ cd ..

That's "cd" followed by a space, and then two dots. This takes us back one directory. We also have another command below:

$ cd ../..

This command takes us back two directories if you happen to be two subdirectories in. So if you're still in the "etc" directory, go back one directory. There are a number of other directories to familiarize with within the root directory. Below are some of those directories:

- /bin – this directory is for commands and binaries.

- /sbin – this is for system and administrator commands.

- /user – this is for user-related files. It also has its own /bin and /sbin subdirectories.

- /var – var stands for variable. The /var/log subdirectory is a common destination for users who troubleshoot problems whose log file end up here.

- /home – As what we discussed before, this is for the user directories.

- /mnt – mnt, which stands for Mount, is pretty much the mount point for any external drives, image files, or things like that.

- /dev – this is for devices that interface with the hardware.

- /lib – this is for programs and program libraries.

There a lot more, but this is just a quick rundown of the more common directories that you're likely to encounter and use.

So how do we add, remove, or modify files and directories? Well, to create a new directory, type in the command below:

$ mkdir directoryname

Mkdir stands for make directory. This should be followed by a space and then the name of the directory that you want to create. To remove that empty directory, just used the command below:

$ rmdir directoryname

Rmdir stands for remove directory. This should be also followed by a space and then the name of the directory that you want to remove. Keep in mind that using this command with a non-empty directory would result to an error. There's another option to remove a directory even though it contains files and subdirectories. However, it can be a little dangerous in the wrong hands if they don't know what they're doing.

So please use with caution whenever you utilize the below command:

$ rm -rf directoryname

The -r option after the remove command means to do the removing process recursively, which drills down into the subdirectories located within the directory you're trying to remove. The -f option stands for Force, which forces the process to go through. So executing this command basically tells the system that you know what you're doing.

The system will allow you to execute the command as long as you have permission to that file or directory. Again, use these options only if you really need to. There won't be any dialogue box, pop-up window, or anything like that to save you from making a mistake with this command.

For copying files and directories, you'll be using the CP command. Look at the example command below:

$ cp filename1 filename2

First type in the cp command, followed by a space, followed by the name of the file to copy from, followed by another space, and then lastly the name of the file to copy to. Note that filename 1 and filename 2 should be different. Also, keep in mind that you need to have root (sudo) access in order to successfully execute this file.

To rename or move a file without duplicating it, use the command below:

$ mv filename1 filename2

The above command will essentially move the contents of filename1 to filename2. You can also specify directories where files are located. For example, let's say you want to copy the contents of /etc/hosts to your home folder. For that you would use the below command:

    $ cp /etc/hosts /home/pi

Type in the cp command, followed by a space, then followed by the file-path of the file that you want to copy, and then another space, and then followed lastly by the file-path of the location where you want to copy the file to.

To move or copy a directory and its subdirectories, you want to use the -r option, which is again a recursive option.

    $ cp -r directory1 directory2

If you want to find out about any of these commands, you can use the MAN command. MAN stands for manual. If you want to find out more about the cp command for example, you just have to type the command below:

    $ man cp

This will pull up all the information about the cp command and all the options associated with it. The same applies to the mv, rm, and so forth.

Chapter 6: Python Programming in Raspberry Pi

In this chapter, we're going to look at the best programming language for the Raspberry Pi; Python. It is considered by many as a notably powerful and dynamic language that is used to develop numerous application domains. Most programmers compare Python to Perl, Java, Ruby, Tcl, or Scheme. This is why Python is the primary language the Raspberry Pi is designed to operate on.

Below are some of the key features of the Python programming language:

- Superb for programming veterans and also remain simple enough for beginners to understand.

- Excellent scalability. Python is excellent in small projects; even better for large scale ones.

- Extremely portable and is compatible across all platforms

- Hardware embeddable

- Elegant and simple to understand syntax and semantics

- Excellent stability

- Has a huge standard library of pre-built subroutines, codes, etc.

Moving forward, we'll discuss how to get a very simple python program working. In addition, we'll also discuss how to use the Graphical User Interface to code basic python programs. However, what this chapter won't tackle is the nitty-gritty of learning how to program in Python. Python object-oriented programming is a whole topic in itself and is therefore out of the scope of this book.

Python programming in the Raspberry Pi is similar to Python programming using a normal desktop or laptop computer. It's just that with the Raspberry Pi, you're programming in a very portable, cheap, but robust platform. To start programming with Python on your Raspberry Pi, you must first make sure that you've successfully installed the Raspbian operating system.

Using the desktop of your Raspbian OS, click on the Menu option on the taskbar. Clicking on menu allows you to see the sub-menus for Raspbian. Click on the "Programming" sub-menu. Once you do that, you'll be presented with different programming applications for Raspbian. One of those applications would be your Python Editor.

Python Editors vary depending on the version of Raspbian running in your Raspberry Pi. However, the most common Python IDE would be Stan's Python Editor or SPE. Click on that and you'll be presented with GUI (Graphical User Interface) for your Python IDE (Integrated Development Environment).

The main window of your Python IDE is where you'll type the source code of the program that you're developing. To start making our simple Python program, first create a folder on your Raspbian desktop where we will put our python script. Just right click on any empty space on your desktop and click "Create New Folder." Don't forget to give your new folder a name.

Now, let's go ahead and develop a simple "Hello World" program using Python. To do so, go to   the code editor window of your Python IDE and type in the following code:

```
1       print "Hello World"
```

After typing this code in your editor, save it by clicking on the "File" menu at the menu of your Python IDE and then click on "Save." You will have to give your python script a name. In this case, just type in the filename as HelloWorld.py and save it inside the new folder that you have created on your desktop. Take note that filenames should not contain any spaces and that it should have a .py extension. This is the file extension for all Python scripts.

After saving the file, run the program by clicking on the "run" button on the menu bar of your Python IDE. At this point, two things should happen depending on which Python IDE you're using. Some IDEs will pull up the terminal and display the output of the program from there, while other IDEs will have a status bar at the bottom where it will show you the output of the Python program that you just ran.

Regardless of whatever mode of output your IDE uses, it should display the words "Hello World." If you want to run your Python script from the terminal itself and not from the IDE, all you need to do is navigate to where your Python script is located first and run the script from there. To do this, type the command below:

$ cd /home/pi/Desktop/foldername

The "foldername" is the name of the folder that you have created on the desktop. Once you're there type the below command to display the contents of that folder:

$ ls -l

Now, you'll see your HelloWorld.py file. To run this just type in the below command:

$ python HelloWorld.py

The command above is basically telling the system to use the Python interpreter to open the HelloWorld.py file. Press ENTER and you'll see the output; "Hello World."

Congratulations, you've just made your first Python program in your Raspberry Pi.

Conclusion

Thank you for purchasing and reading this book. We hope that we've taught you all the basic things that you need to know about your Raspberry Pi; how to set it up, how to install the operating system so that you'll be able to develop programs using it.

From here, developing more intricate programs using Python is just a matter of learning the nitty-gritty of the programming language itself. There are many online tutorials out there that you can go and see to get in-depth knowledge of Python programming. Python is a very versatile and stable programming language and using it to make your Pi do wondrous things is easy, as long as you have an advanced knowledge of the programming language itself.

Again, Thank you so much and we hope that you succeed in your quest for knowledge on the Python programming language and that you have fun in finding ingenious ways to use your Raspberry Pi.